# Basic Materials

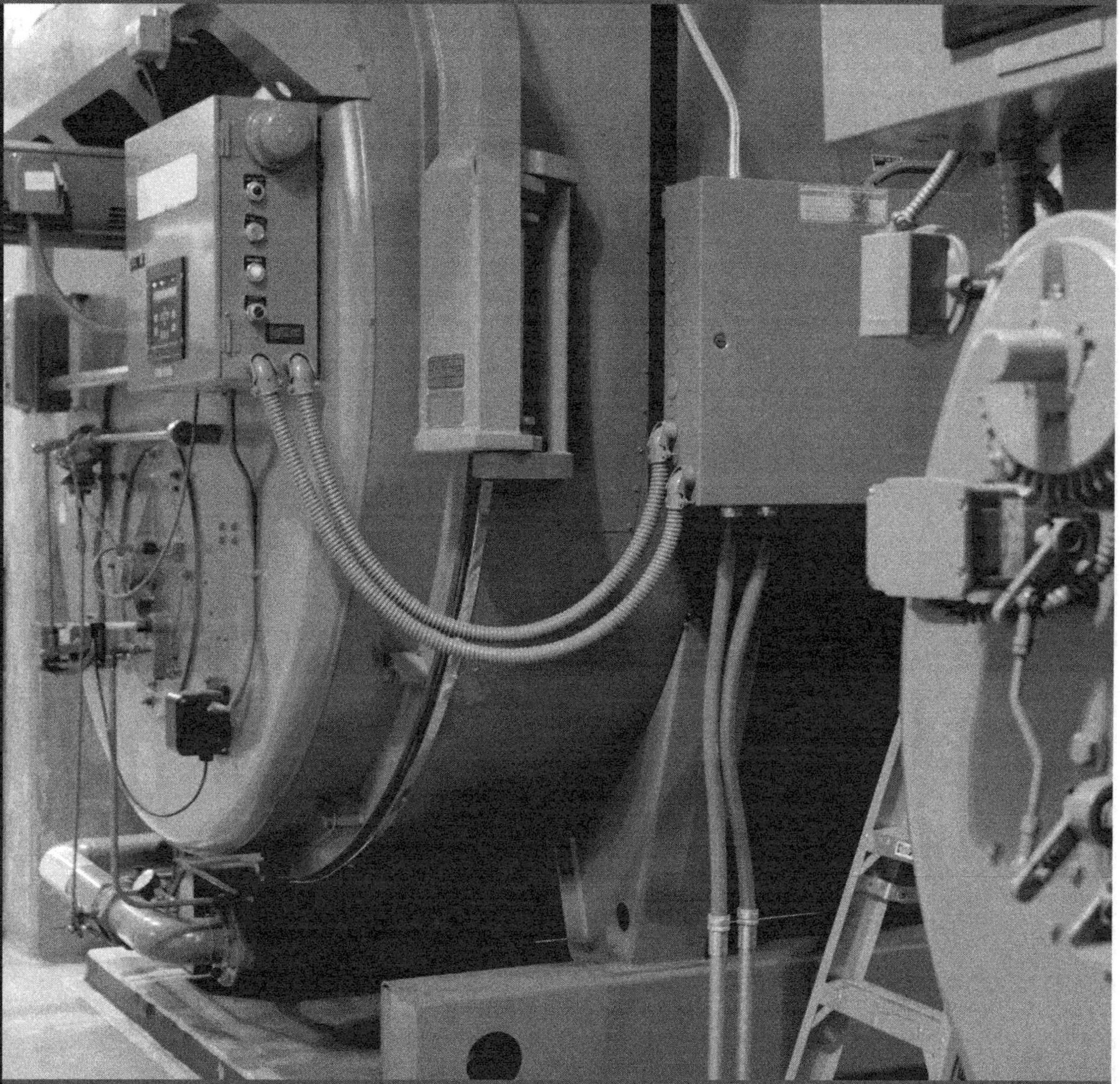

34104-10

## BOILERMAKING LEVEL ONE

34108-10
**Welding Basics**

34107-10
**Base Metal Preparation**

34106-10
**Cutting and Fitting Gaskets**

34105-10
**Oxyfuel Cutting**

34104-10
**Basic Materials**

34103-10
**Boilermaking Tools**

34102-10
**Boilermaking Safety**

34101-10
**Introduction to Boilermaking**

**Core Curriculum:
Introductory Craft Skills**

This course map shows all of the modules in *Boilermakling Level One*. The suggested training order begins at the bottom and proceeds up. Skill levels increase as you advance on the course map. The local Training Program Sponsor may adjust the training order.

## Objectives

When you have completed this module, you will be able to do the following:

1. Describe materials used in boiler construction and explain where these materials are used.
2. Describe the different types of iron and steel.
3. Identify codes and markings used in material identification.
4. Describe material properties of the refractory, insulation, and ceramic material used in boiler construction.

## Trade Terms

American Iron and Steel Institute (AISI)
Alloy
ASTM International
American Society of Mechanical Engineers (ASME)
Coefficient of thermal expansion
Ductility

Ferrous
Malleability
SAE International
Steel
Unified Numbering System (UNS)
Wrought
Yield stress

## Prerequisites

Before you begin this module, it is recommended that you successfully complete *Core Curriculum* and *Boilermaking Level One*, Modules 34101-10 through 34103-10.

# Contents

*Topics to be presented in this module include:*

# 1.0.0 INTRODUCTION

Boilers are built to withstand extreme pressures and temperatures. To meet these conditions, special materials are used in their construction.

Boilermakers must understand these materials, their physical properties, where they are used, and how to identify them. This module introduces the trainee to the materials used in boiler construction.

# 2.0.0 IMPORTANT MATERIAL PHYSICAL PROPERTIES

All materials react differently under given conditions. Temperature affects some materials more than others. Some materials resist flame. Some materials weigh very little, yet can withstand thousands of pounds of pressure. It is important to understand physical properties so that the right material can be chosen for a given location during boiler construction.

The following sections introduce the trainee to the physical properties that must be considered when selecting materials for use in boiler construction.

## 2.1.0 Chemical Composition

The chemical composition of a metal or other material is the property that determines all the other physical properties of that material.

### 2.1.1 Elements

An element is a pure substance made of just one kind of matter. Everything is made of one or more elements. To keep track of the elements, scientists devised a chart called the periodic table of elements. This table groups elements with like properties together, and gives other information about each element. *Figure 1* shows a periodic table of the elements.

### 2.1.2 Compounds

Whenever two or more elements join together chemically they form a compound. Water, for example, is a compound of hydrogen and oxygen. Hydrogen is a highly flammable gas. Oxygen is a gas that supports burning. These two elements, when chemically bonded together, form a liquid that puts fires out.

### 2.1.3 Mixtures

A mixture differs from a compound in that the parts of a mixture are not chemically combined. An example of a mixture is water and salt. These two compounds can be easily mixed, yet they can be easily separated with heat.

### 2.1.4 Alloys

**Alloys** are mixtures of different metals and/or other elements. For example, to make carbon **steel**, iron is heated to very high temperatures so that carbon can be dissolved in and mixed with the iron. As the mixture cools it becomes solid. The result is the alloy carbon steel.

The different elements added to an alloy result in different physical properties. The various physical properties are used to select materials for construction of a boiler.

For example, carbon steel has greater strength (known as **yield stress**) than iron without carbon, at a reasonable cost. Stainless steel is an alloy containing iron and chromium. This alloy provides superior strength. It also provides corrosion resistance. One disadvantage of stainless steel is its cost.

Most alloys used in boiler construction are selected with heat resistance, corrosion resistance, or strength in mind. Costs are also calculated before choosing an alloy for use in boiler construction.

## 2.2.0 Density

Density is the ratio of the mass of a substance to its volume. Density may also be expressed as specific gravity or specific density. Specific gravity is the weight of a material compared to the weight of an equal volume of water. The higher the density (or specific gravity) of a material, the heavier it will be.

Light metals such as aluminum and magnesium have low densities. **Ferrous** metals like carbon steel and stainless steel have much higher densities. An iron or steel item will weigh nearly three times as much as an aluminum item of the same size. *Table 1* lists the densities (weight/volume) of some common materials.

The density of material is often used when estimating the weight of an object before a rigging operation.

## 2.3.0 Heat Transfer Rates

The ability of a material to transfer heat is an important factor when selecting materials for boiler construction. The heat generated by the fuel of the boiler will be wasted if the material used in the boiler tubes does not conduct heat efficiently. Conversely, the materials used to insulate the boiler must not transfer heat easily. The heat generated in the furnace would then be wasted.

The heat transfer capability of a material is known as its thermal conductivity.

## 2.4.0 Thermal Expansion

The amount a material expands when it changes temperature is also an important factor when considering boiler materials. Temperatures vary widely between operating and shutdown conditions. Materials with very high expansion rates can cause extreme stresses to build up in the boiler during these heat-up and cool-down periods. The tendency of a material to expand when it heats up is called the **coefficient of thermal expansion**. Some coefficients are listed in *Table 2*.

## 2.5.0 Melting Point

The melting point of a solid is the temperature at which the solid changes into a liquid. *Table 3* lists the melting points for some common materials. By comparison, the temperature of a light bulb filament is approximately 3,200°F. A common backyard barbecue operates at approximately 350°F.

## 2.6.0 Corrosion Resistance

Corrosion resistance is the ability of a material to resist chemical attack. The metals in a boiler contact many different chemicals. Corrosion resistance is a major reason stainless steel is so common in boiler construction.

## 2.7.0 Mechanical Properties

The mechanical properties of metals determine how they will react to or be affected by the external forces applied to them.

### 2.7.1 Material Strength

Material strength is the mechanical property of a metal. It is measured in many different ways. Some of the common measurements include the following:

- *Tensile strength* – The ability of a metal to withstand tension or a pulling stress. This stress would be found in a bolt as it is tightened.
- *Compressive strength* – The ability of a metal to withstand compression. This stress would be found in the foundation of a heavy structure.
- *Fatigue strength* – The ability of a metal to undergo a repeating stress or load. This stress would be like that found in a spring on a car going down a bumpy road.
- *Flex strength* – The ability of a metal to resist a bending force or stress. This stress would be found in a horizontal support beam.
- *Shear strength* – The ability of a metal to withstand a shearing load. This stress would be found in a horizontal mounting bolt for a heavy piece of equipment.

**PERIODIC TABLE OF THE ELEMENTS**

| Group / Period | 1<br>IA<br>1A | 2<br>IIA<br>2A | 3<br>IIIB<br>3B | 4<br>IVB<br>4B | 5<br>VB<br>5B | 6<br>VIB<br>6B | 7<br>VIIB<br>7B | 8<br>VIIIB<br>8B | 9<br>VIIIB<br>8B | 10<br>VIIIB<br>8B | 11<br>IB<br>1B | 12<br>IIB<br>2B | 13<br>IIIA<br>3A | 14<br>IVA<br>4A | 15<br>VA<br>5A | 16<br>VIA<br>6A | 17<br>VIIA<br>7A | 18<br>VIIIA<br>8A |
|---|---|---|---|---|---|---|---|---|---|---|---|---|---|---|---|---|---|---|
| 1 | 1<br>H<br>1.0080 | | | | | | | | | | | | | | | | | 2<br>He<br>4.00260 |
| 2 | 3<br>Li<br>6.941 | 4<br>Be<br>9.01218 | | | | | | | | | | | 5<br>B<br>10.81 | 6<br>C<br>12.0111 | 7<br>N<br>14.0067 | 8<br>O<br>15.9994 | 9<br>F<br>18.9984 | 10<br>Ne<br>20.179 |
| 3 | 11<br>Na<br>22.9898 | 12<br>Mg<br>24.305 | | | | | | | | | | | 13<br>Al<br>26.9815 | 14<br>Si<br>28.086 | 15<br>P<br>30.9738 | 16<br>S<br>32.06 | 17<br>Cl<br>35.453 | 18<br>Ar<br>39.948 |
| 4 | 19<br>K<br>39.102 | 20<br>Ca<br>40.08 | 21<br>Sc<br>44.956 | 22<br>Ti<br>47.90 | 23<br>V<br>50.941 | 24<br>Cr<br>51.996 | 25<br>Mn<br>54.9380 | 26<br>Fe<br>55.847 | 27<br>Co<br>58.9332 | 28<br>Ni<br>58.71 | 29<br>Cu<br>63.54 | 30<br>Zn<br>65.37 | 31<br>Ga<br>69.72 | 32<br>Ge<br>72.59 | 33<br>As<br>74.9216 | 34<br>Se<br>78.96 | 35<br>Br<br>79.904 | 36<br>Kr<br>83.80 |
| 5 | 37<br>Rb<br>85.4678 | 38<br>Sr<br>87.62 | 39<br>Y<br>88.906 | 40<br>Zr<br>91.22 | 41<br>Nb<br>92.906 | 42<br>Mo<br>95.94 | 43<br>Tc<br>99 | 44<br>Ru<br>101.07 | 45<br>Rh<br>102.906 | 46<br>Pd<br>106.4 | 47<br>Ag<br>107.870 | 48<br>Cd<br>112.40 | 49<br>In<br>114.82 | 50<br>Sn<br>118.69 | 51<br>Sb<br>121.75 | 52<br>Te<br>127.60 | 53<br>I<br>126.9045 | 54<br>Xe<br>131.30 |
| 6 | 55<br>Cs<br>132.91 | 56<br>Ba<br>137.34 | 57-71<br>Rare Earths | 72<br>Hf<br>178.49 | 73<br>Ta<br>180.95 | 74<br>W<br>183.8 | 75<br>Re<br>186.2 | 76<br>Os<br>190.2 | 77<br>Ir<br>192.2 | 78<br>Pt<br>195.09 | 79<br>Au<br>196.967 | 80<br>Hg<br>200.59 | 81<br>Tl<br>204.37 | 82<br>Pb<br>207.2 | 83<br>Bi<br>208.9806 | 84<br>Pa<br>210 | 85<br>At<br>210 | 86<br>Rn<br>222 |
| 7 | 87<br>Fr<br>223 | 88<br>Ra<br>226.03 | 89-103<br>Radioactive Rare Earths | 104<br>Rf<br>261 | 105<br>Ha<br>262 | 106<br>Sg<br>263 | 107<br>Bh<br>264 | 108<br>Hs<br>265 | 109<br>Mt<br>266 | 110<br>Uun<br>271.15 | 111<br>Uuu<br>272.15 | 112<br>Uub<br>277 | 113<br>Uut<br>? | 114<br>Uuq<br>285 | 115<br>Uup<br>? | 116<br>Uub<br>289 | 117<br>Uus<br>? | 118<br>Uuo<br>? |

Representative Elements: groups 1, 2 and groups 13–18. Transition Elements: groups 3–12.

104F01.EPS

*Figure 1* Periodic table of the elements.

These strengths are all related to the microscopic structure of a metal or alloy. They are also closely related to the following mechanical properties:

**Table 1** Densities of Common Materials

| Material | Density (Lbs/Cubic Foot) |
|---|---|
| Water | 62 |
| Cement | 144 |
| Granite | 159 |
| Common glass | 162 |
| Aluminum | 165 |
| Zinc | 440 |
| Iron or steel | 485 |
| Copper | 556 |
| Lead | 710 |
| Tungsten | 1,200 |
| Gold | 1,205 |

104T01.EPS

**Table 2** Coefficients of Expansion for Common Materials

| Material | Coefficient of Expansion* |
|---|---|
| Glass | 0.0000050 |
| Stainless steel | 0.0000060 |
| Mild steel | 0.0000073 |
| Ductile iron | 0.0000063 |
| Inconel® | 0.0000070 |
| Copper | 0.0000093 |
| Silver | 0.0000106 |
| Bronze | 0.0000100 |
| Lead | 0.0000151 |
| PVC | 0.0000280 |

*Inches

104T02.EPS

**Table 3** Melting Points of Some Common Materials

| Material | Melting Point (°F) |
|---|---|
| Zinc | 787 |
| Magnesium | 1,202 |
| Aluminum | 1,215 |
| Glass | 1,500 |
| Copper | 1,981 |
| Manganese | 2,273 |
| Nickel | 2,651 |
| Iron | 2,800 |
| Titanium | 3,300 |
| Molybdenum | 4,760 |

104T03.EPS

- *Hardness* – The measure of a metal's resistance to penetration or deformation.
- *Brittleness* – The property of a material that causes it to shatter when bent or deformed.
- **Ductility** – The ability of a metal to stretch before it breaks.
- *Wear resistance* – The ability of a metal to resist abrasion or wear.
- *Toughness* – The ability of a metal to resist shock or sudden impact.

An ideal metal would have all of these properties, but actual metals have limitations. The application for the metal must be considered before choosing the right alloy for use in a boiler. The stresses it will undergo must also be considered.

# 3.0.0 METALS USED IN THE CONSTRUCTION OF BOILERS

Metals are classified into two basic groups: ferrous metals, which contain mainly iron; and nonferrous metals, which contain very little or no iron. Ferrous comes from the Latin word for iron, *ferrum*. This is why the symbol for iron on the periodic table is Fe. Ferrous metals include all of the steels, cast irons, **wrought** irons, malleable irons, and ductile (nodular) irons. Nonferrous metals include the light metals (aluminum, magnesium, titanium) and their alloys, the heavy metals (copper, nickel, lead, tin, zinc), and the precious metals (platinum, gold, silver) and their alloys.

## 3.1.0 Ferrous Metals

Ferrous metals contain mostly iron and some carbon. The largest group of the ferrous-based metals is known as carbon steel. As carbon content increases in carbon steels, the steel becomes stronger and more wear resistant. The carbon content of a steel is the largest factor in determining its use. Steel castings usually contain a higher percentage of carbon than rolled plate and other rolled shapes.

**SAE International** is the standards organization that governs the marking codes for steels commonly used in structural shapes, plate, strip, sheet, and welded tubing. Steel alloys are identified by a four- or five-digit code. The marking codes may be referred to as the SAE system. The number designation is described in *Figure 2*.

The type of steel is determined by the amount of carbon and the other metals in the mixture. The 40 indicates the alloy content (0.25 percent molybdenum) and the 10 indicates the carbon content (0.10 percent carbon). *Table 4* lists the common steel alloy content codes.

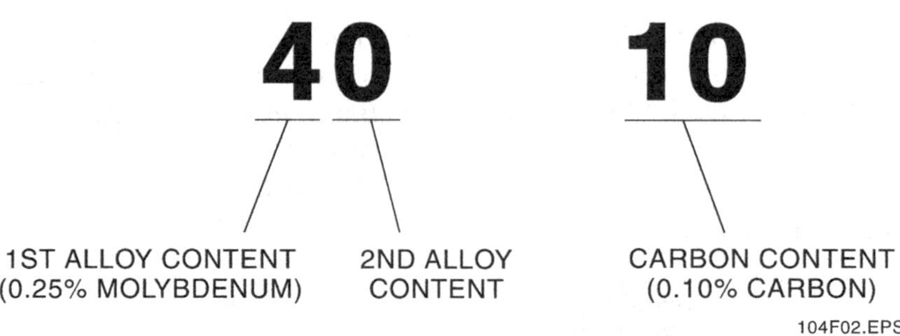



*Figure 2* Steel designation example.

**Table 4** Steel Alloy Content Codes

| SAE Number | Carbon Steel Types and Classes |
|---|---|
| 1XXX | Carbon steels |
| 10XX | Plain (nonresulfurized) carbon steel grades |
| 11XX | Free machining, resulfurized (screw stock) |
| 12XX | Free machining, resulfurized, rephosphorized |
| 13XX | Manganese 1.75% |
| 15XX | High manganese carburizing steels |
| 2XXX | Nickel steels |
| 23XX | 3.50% nickel |
| 25XX | 5.00% nickel |
| 3XXX | Nickel-chromium steels |
| 31XX | 1.25% nickel, 0.65% or 0.80% chromium |
| 32XX | 1.75% nickel, 1.00% chromium |
| 33XX | 3.50% nickel, 1.50% chromium |
| 30XX | Corrosion and heat-resisting steels |
| 4XXX | Molybdenum steels |
| 40XX | 0.25% molybdenum |
| 41XX | 0.50% to 0.95% chromium, 0.12% or 0.20% molybdenum |
| 43XX | 1.80% nickel, 0.50% or 0.80% chromium, 0.25% molybdenum |
| 46XX | 1.55% or 1.80% nickel, 0.20% or 0.25% molybdenum |
| 47XX | 1.05% nickel, 0.45% chromium, 0.25% molybdenum |
| 48XX | 3.50% nickel, 0.25% molybdenum |
| 50XX | 0.28% or 0.40% chromium |
| 51XX | 0.80%, 0.90%, 0.95%, 1.0% or 1.05% chromium |
| 5XXXX | 1.00% carbon, 0.50%, 1.00% or 1.45% chromium |
| 6XXX | Chromium vanadium steels |
| 61XX | 0.80% or 0.95% chromium, 0.10% (minimum) vanadium |
| 86XX | 0.55% nickel, 0.50% or 0.65% chromium, 0.20% molybdenum |
| 92XX | 0.85% manganese, 2.00% silicon |
| 93XX | 3.25% nickel, 1.20% chromium, 0.12% molybdenum |
| 94XX | 1.00% manganese, 0.45% nickel, 0.40% chromium, 0.12% molybdenum |
| 97XX | 0.55% nickel, 0.17% chromium, 0.20% molybdenum |
| 98XX | 1.00% nickel, 0.80% chromium, 0.25% molybdenum |
| XXBXX | Boron steels (0.0005% minimum boron) |

104T04.EPS

The carbon content of steel ranges from 0.1 to 2.0 percent. The lower the carbon content, the softer the steel. The higher the carbon content, the harder the steel. Above 2.0 percent carbon the alloy is no longer considered steel, but cast iron. Wrought iron contains essentially zero carbon. The following examples explain the SAE numbers for carbon steels.

*SAE Number 1020:*
- 10  Designates carbon steel, non-resulfurized
- 20  Contains approximately 0.20 percent carbon

*SAE Number 2512:*
- 25  Designates steel alloyed with approximately 5 percent nickel
- 12  Designates steel containing approximately 0.12 percent carbon

*SAE Number 52100:*
- 52  Contains 1.45 percent chromium
- 100  Designates approximately 1 percent carbon

## 3.2.0 Steel and Steel Alloys

All steels contain impurities and trace amounts of elements such as manganese, phosphorus, silicon, and sulfur. Sometimes additional amounts of these elements are added to steel.

The effect of adding elements to steel vary greatly. Different steel alloys are used in different parts of the boiler.

Alloy steels are steels that contain significant amounts of one or more of the following elements:

- *Manganese (Mn)* – Promotes **malleability**
- *Silicon (Si)* – Prevents the formation of blow-holes
- *Copper (Cu)* – Provides corrosion resistance
- *Boron (B)* – Increases hardness
- *Chromium (Cr)* – Increases strength and corrosion resistance
- *Cobalt (Co)* – Increases hardness and resistance to wear
- *Molybdenum (Mo)* – Increases high-temperature strength and hardness
- *Nickel (Ni)* – Increases strength and toughness
- *Tungsten (W)* – Increases high-temperature strength and wear resistance
- *Vanadium (V)* – Increases toughness

### 3.2.1 Stainless Steel Classification

Stainless steels are corrosion-resisting steels. They are alloys of iron with chromium, or with chromium and nickel, or with nickel. Often other alloying metals are added to enhance physical and mechanical properties. Stainless steels do not rust. They are resistant to attack by various chemicals. Many have good low- and high-temperature mechanical characteristics. Compared with mild steels, stainless steels have the following:

- Lower melting points
- Lower coefficients of thermal conductivity, which decreases the chance of distortion
- Higher coefficients of thermal expansion, which increases the chance of distortion
- Higher electrical resistances, which increases the build-up of heat from welding current

The stainless steels are grouped into five basic types:

- *Chromium-Nickel-Magnesium-Austenitic* – Nonmagnetic in the annealed condition and non-hardenable by heat treatment
- *Chromium-Nickel-Austenitic* – Nonmagnetic in the annealed condition and non-hardenable by heat treatment
- *Chromium-Martensitic* – Magnetic and hardenable by heat treatment
- *Chromium-Ferritic* – Magnetic and non-hardenable by heat treatment
- *Martensitic* – Magnetic and hardenable by heat treatment

The **American Iron and Steel Institute (AISI)** Stainless Steel Classification System uses three digits (with some suffixes) to designate stainless steels (*Table 5*). The first digit indicates the basic type. The last two digits indicate the specific alloy.

The austenitic stainless steels contain both chromium and nickel. Austenitic stainless steels are the types most often encountered. The higher carbon content of these steels tends to produce weld boundaries of depleted chromium. This lowers corrosion resistance. It also increases crack sensitivity in the weld boundary zone. This occurs because precipitated carbon unites with the chromium in the weld boundary areas. The carbon and chromium mix to form carbides of chromium. To prevent boundary depletion, low-carbon, high-chromium content filler metals should be used.

The 4XX series of stainless steels contain only chromium and small amounts of other elements. They are known as straight chrome types. They are highly magnetic.

| AISI Number | C% | Mn% | Si% | Cr% | Ni% | Other Elements |
|---|---|---|---|---|---|---|
| **Chromium-Nickel-Magnesium-Austenitic (Non-Hardenable)** | | | | | | |
| 201 | 0.15 max | 5.5/7.5 | 1.0 | 16.0–18.0 | 3.5/5.5 | $N_2$ 0.25 Max |
| 202 | 0.15 max | 7.5/10 | 1.0 | 17.0–19.0 | 4.0/6.0 | $N_2$ 0.25 Max |
| **Chromium-Nickel-Austenitic (Non-Hardenable)** | | | | | | |
| 301 | 0.15 max | 2.0 | 1.0 | 16.0–18.0 | 6.0/8.0 | — |
| 302 | 0.15 max | 2.0 | 1.0 | 17.0–19.0 | 6.0/10.0 | — |
| 302B | 0.15 max | 2.0 | 2.0/3/0 | 17.0–19.0 | 6.0/10.0 | — |
| 303 | 0.15 max | 2.0 | 1.0 | 17.0–19.0 | 6.0/10.0 | S 0.15 min |
| 303Se | 0.15 max | 2.0 | 1.0 | 17.0–19.0 | 6.0/10.0 | Se 0.15 min |
| 304 | 0.08 max | 2.0 | 1.0 | 18.0–20.0 | 8.0/12.0 | — |
| 304L | 0.03 max | 2.0 | 1.0 | 18.0–20.0 | 8.0/12.0 | — |
| 305 | 0.12 max | 2.0 | 1.0 | 17.0–19.0 | 10.0/13.0 | — |
| 308 | 0.08 max | 2.0 | 1.0 | 19.0–21.0 | 10.0/12.0 | — |
| 309 | 0.20 max | 2.0 | 1.0 | 22.0–24.0 | 12.0/15.0 | — |
| 309S | 0.08 max | 2.0 | 1.0 | 22.0–24.0 | 12.0/15.0 | — |
| 310 | 0.25 max | 2.0 | 1.5 | 24.0–26.0 | 19.0/22.0 | — |
| 310S | 0.08 max | 2.0 | 1.5 | 24.0–26.0 | 19.0/22.0 | — |
| 314 | 0.25 max | 2.0 | 1.5/3.0 | 23.0–26.0 | 19.0/22.0 | — |
| 316 | 0.08 max | 2.0 | 1.0 | 16.0–18.0 | 10.0/14.0 | Mo 2.0/3.0 |
| 316L | 0.03 max | 2.0 | 1.0 | 16.0–18.0 | 10.0/14.0 | Mo 2.0/3/0 |
| 317 | 0.08 max | 2.0 | 1.0 | 18.0–20.0 | 11.0/15.0 | Mo 3.0/4.0 |
| 321 | 0.08 max | 2.0 | 1.0 | 17.0–19.0 | 9.0/12.0 | Ti 5xC min |
| 347 | 0.08 max | 2.0 | 1.0 | 17.0–19.0 | 9.0/13.0 | Cb+Ta10xCmin |
| 348 | 0.08 max | 2.0 | 1.0 | 17.0–19.0 | 9.0/12.0 | Ti 0.10 max |
| **Chromium-Martensitic (Hardenable)** | | | | | | |
| 403 | 0.15 max | 2.0 | 1.0 | 11.5–13.0 | — | — |
| 410 | 0.15 max | 2.0 | 1.0 | 11.5–13.0 | — | — |
| 414 | 0.15 max | 1.0 | 1.0 | 11.5–13.0 | 1.25/2.5 | — |
| 416 | 0.15 max | 1.25 | 1.0 | 12.0–14.0 | — | S 0.15 min |
| 416Se | 0.15 max | 1.25 | 1.0 | 12.0–14.0 | — | Se 0.15 min |
| 420 | 0.16 max | 1.0 | 1.0 | 12.0–14.0 | — | — |
| 430 | 0.20 max | 1.0 | 1.0 | 15.0–17.0 | 1.25/2.5 | — |
| 440A | 0.60/0.75 | 1.0 | 1.0 | 16.0–18.0 | — | Mo 0.75 max |
| 440B | 0.75/0.95 | 1.0 | 1.0 | 16.0–18.0 | — | Mo 0.75 max |
| 440C | 0.95/1.2 | 1.0 | 1.0 | 16.0–18.0 | — | Mo 0.75 max |
| **Chromium-Ferritic (Non-Hardenable)** | | | | | | |
| 405 | 0.08 max | 1.0 | 0.5 | 11.5–14.5 | — | Al 1.1/0.3 |
| 430 | 0.12 max | 1.0 | 1.0 | 14.0–18.0 | — | — |
| 430F | 0.12 max | 1.25 | 1.0 | 14.0–18.0 | — | S 0.15 min |
| 430FSe | 0.12 max | 1.25 | 1.0 | 14.0–18.0 | — | Se 0.15 min |
| 446 | 0.20 max | 1.50 | 1.0 | 23.0–27.0 | — | N 0.25 max |
| **Martensitic** | | | | | | |
| 501 | 0.11 max | 1.0 | 1.0 | 4.0/6.0 | — | Mo 0.4/0.65 |
| 502 | 0.10 max | 1.0 | 1.0 | 4.0/6.0 | — | Mo 0.4/0.65 |

104T05.EPS

### 3.2.2 Unified Numbering System

The **Unified Numbering System (UNS),** short for Unified Numbering System of Metals and Alloys, is a designation system used in North America only for commercial metals and alloys that are in active use. The UNS provides a means of linking many nationally used numbering systems currently administered by societies, trade associations, and individual users and producers of metals and alloys. This avoids confusion caused by use of more than one identification number for the same material. It also avoids confusion caused by having the same number assigned to two or more entirely different materials.

The UNS number includes a prefix letter and five digits designating a specific material composition. The prefix letter indicates the type of alloy. A portion of the five digits may match older three- or four-digit numbering systems. The remaining digits indicate more modern variations. The UNS is managed jointly by **ASTM International** and SAE International, and is defined in *ASTM Standard E527-07*.

The UNS number alone does not represent a material specification. It establishes no requirements for material properties, heat treatment, form, or quality. It is only used to identify metals and alloys that have controlling limits established in specifications published elsewhere. The UNS number can be used to index the controlling specifications for a specific metal or alloy. The following is a summary of the types of metals and alloys covered for each of the prefix letters:

- *Axxxxx* – Aluminum and aluminum alloys
- *Cxxxxx* – Copper and copper alloys
- *Dxxxxx* – Specified mechanical property steels
- *Exxxxx* – Rare earth and rare earthlike metals and alloys
- *Fxxxxx* – Cast irons
- *Gxxxxx* – SAE carbon and alloy steels (except tool steels)
- *Hxxxxx* – SAE H-steels
- *Jxxxxx* – Cast steels (except tool steels)
- *Kxxxxx* – Miscellaneous steels and ferrous alloys
- *Lxxxxx* – Low-melting metals and alloys
- *Mxxxxx* – Miscellaneous nonferrous metals and alloys
- *Nxxxxx* – Nickel and nickel alloys
- *Pxxxxx* – Precious metals and alloys
- *Rxxxxx* – Reactive and refractory metals and alloys
- *Sxxxxx* – Heat- and corrosion-resistant (stainless) steels
- *Txxxxx* – Tool steels, wrought and cast
- *Wxxxx* – Welding filler metals
- *Zxxxxx* – Zinc and zinc alloys

### 3.3.0 Heat Number

When steel and steel alloys are manufactured, the individual ingredients are mixed together in a furnace. Each batch is heated and mixed individually. The ingredients of the mixture or alloy are carefully controlled and recorded. This record follows the batch as it is formed, cooled, and heat-treated.

When the manufacturing process is complete, the final product is assigned a six- or seven-digit alphanumeric code. This is known as its heat number (*Figure 3*). This number identifies the batch, no matter how the product is machined or formed in the future. The heat number can identify the specific batch this metal was poured from. It can also identify the heat treatment process it was subjected to.

Each manufacturer has a different format for these numbers. Records for each batch are maintained by the manufacturer. The heat number must stay with the metal product until final installation.

The heat number must be on any piece of steel or pipe used in the pressure components of a boiler. If a piece is to be cut off a length of stock, the heat number must be transferred to the cut piece before cutting. This is done by hand stamping or printing. It can also be transferred by an attached embossed or engraved metal tag. If the number is not transferred, an inspector can reject an installation. In some cases, the applicable **American Society of Mechanical Engineers (ASME)** and/or ASTM International codes and other information covered in the next section may have to be transferred. *Figure 4* shows examples of equipment that can be used for heat number and code number transfer.

HEAT NUMBER

104F03.EPS

*Figure 3* A stamped heat number.

### 3.4.0 Standardization for the Boilermaker Trade

Many organizations set standards for trades or industries. For construction and repair of boilers, ASME sets the standards through the *ASME Boiler and Pressure Vessel Code*. Other organizations setting standards that may affect boilermaking include the following:

- *AISC* – American Institute of Steel Construction
- *AISI* – American Iron and Steel Institute
- *ASTM International* – Originally known as American Society for Testing and Materials
- *ANSI* – American National Standards Institute
- *API* – American Petroleum Institute
- *SAE International* – Originally known as Society of Automotive Engineers
- *ISO* – International Standards Organization
- *CSA* – Canadian Standards Association

These and other organizations set standards for industry to follow. The standard-setting organizations are made up of representatives from industry. These organizations are not part of any government agency.

### 3.5.0 Markings/Identification

Standards for the marking and identification of materials used in the construction of boilers are contained in several general code sections for the various components in both the *ASME Boiler and Pressure Vessel Code* and the *ASTM Code, Volume One*. These two codes follow each other very closely. Additional instructions for marking are listed in each section of the ASME code. Markings are generally painted or inked onto the material. There are exceptions. Sometimes material is identified by one tag for an entire pallet; sometimes the markings are stamped on the material. The markings contain the following information:

- Manufacturer's name
- ASTM XXX (the letters ASTM followed by the applicable code for the material being marked)
- An ASME code
- Grade X (the word *grade* followed by the alphanumeric designation for the grade, usually A, B, C, T2, or other; grades are specified in individual codes)

Other markings that may be found, depending on the application or code, include the following:

- HOT FINISHED
- COLD DRAWN
- Buyer name and order number
- Heat number
- Heat treatment lot number
- S (seamless)
- ERW (electric resistance welded)
- Not Annealed
- LT XXX (low temperature followed by the impact test temperature)
- Material code (examples: CMS – chromium, magnesium, silicon; MO – molybdenum; MV – manganese, vanadium)
- X, Y, or Z (codes indicating that the manufacturer did not perform an ASME-required test on the material)

ASME code formats are two letters, a dash, and then two to three numbers. For example:

- SA codes are all ferrous material
- SB codes are all nonferrous material
- SA–450 is the code for general requirements for ferrous tubing
- SB–450 is the code for general requirements for nonferrous tubing

### 4.0.0 STRUCTURAL STEEL AND COMMON MILLED SHAPES

Metals are commercially available in many shapes and sizes. Metals may be shaped by casting, rolling, drawing, forging, cutting, machining, welding, and spinning. Structural steel includes a wide range of steel types, classes, and shapes. Structural steel and other common milled shapes are formed by rolling hot or cold metal between a succession of rollers. The rollers are specially configured for the shapes they are to produce.

To ensure uniform standards, ASTM International specifies the properties of strength, weight, corrosion resistance, and weldability for various steel classifications. *Table 6* lists common types and classifications of structural steel. The column labeled Minimum Yield Stress is a measure of the strength of the material.

### 4.1.0 Structural Steel Classifications

The shape or type of structural steel is identified on drawings and the bill of lading with a symbol or abbreviation. Size and dimensions are always given in a specified order. The specification format for designating shape and size is shown with the relevant shape in the following sections.

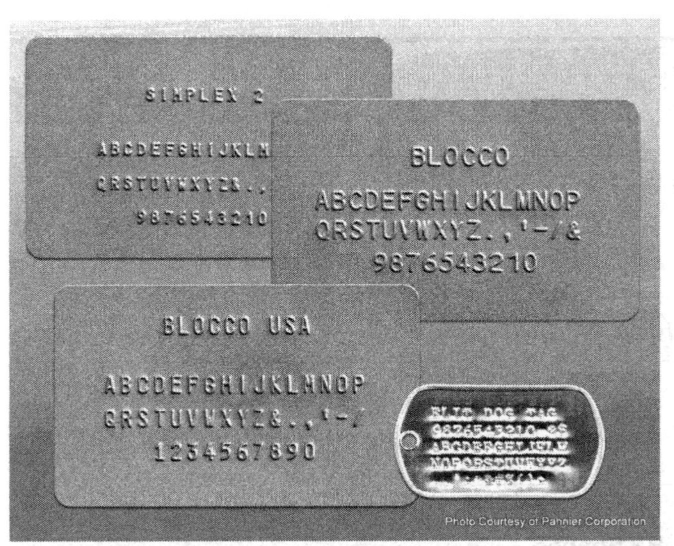

METAL TAG EMBOSSER

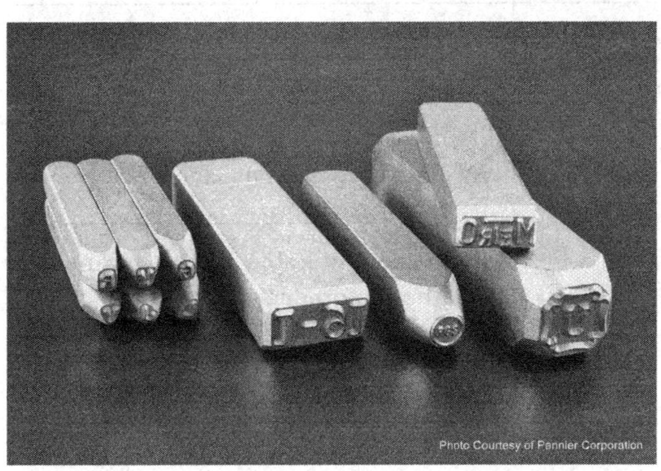
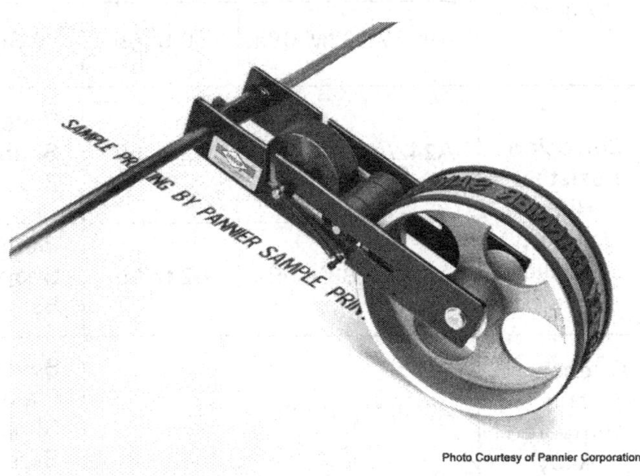


HAND STAMPS

HAND WHEEL CODE PRINTER

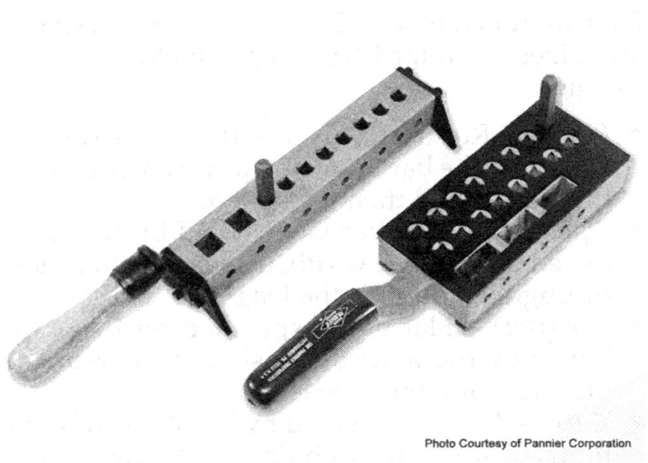
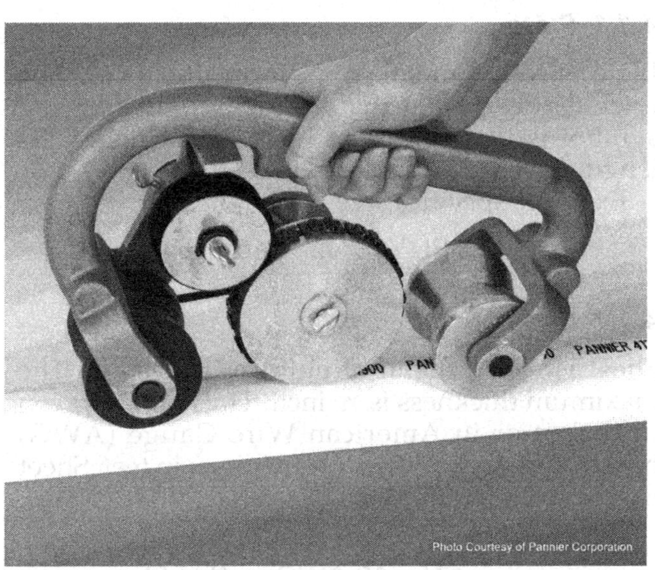


HAND STAMP HOLDER

PIPE, TUBE, AND BAR PRINTER

104F04.EPS

*Figure 4* Examples of hand stamping, printing, and embossing equipment.

**Table 6** Structural Steel Classifications

| Steel Type | ASTM Class Standard | Minimum Yield Stress | Form | Remarks |
|---|---|---|---|---|
| Carbon Steel | A36/A36M-08 | 36 | Plates<br>S, M, & HP Shapes<br>Angles<br>Bars<br>Sheets, Strips<br>Rivets<br>Bolts, Nuts | For buildings and general structures available in high toughness grades |
| | A529/A529-05 | 42 | Plates<br>Shapes<br>Bars | For buildings and similar construction |
| High Strength Low Alloy | A572/A572M-07 | 40 to 65 | Plates<br>S, M, & HP Shapes<br>Bars | Primarily for lightweight buildings and bridges |
| | A992/A992M-06a | 50 to 65 | W-Shapes | Lightweight high toughness for buildings |
| Corrosion Resistant High Strength Low Alloy | A242/A242M-04 | 42 to 50 | Plates<br>Shapes<br>Bars | Lightweight and added durability<br>Weathering grades available |
| | A588/A588M-05 | 42 to 50 | Plates<br>Shapes<br>Bars | Lightweight, durable in high thickness<br>Weathering grades available |
| Quenched and Tempered Alloy | A514/A514M-05 | 90 to 100 | Several Types<br>Plates<br>Shapes<br>Bars | Stength varies with thickness and type |

104T06.EPS

### 4.1.1 Plate

Plate is rolled metal of uniform thickness. The thickness is greater than $\frac{3}{16}$ inch. Its identifying symbol is PL. Thickness is given in inches. Its length and width are given in feet.

Example specification:

PL ¼" × 4' × 10'

### 4.1.2 Sheet

Sheet is rolled metal of uniform thickness. The maximum thickness is $\frac{3}{16}$ inch. Thickness is given in inches or in American Wire Gauge (AWG) number. Length and width are given in feet. Sheet is often rolled into coils.

Example specification:

Sheet No. 12 AWG × 4' × 8'

### 4.1.3 Bars

Bars are rolled to a variety of cross section shapes and sizes. Standard bar shapes include the following:

- *Round* – Round bar is specified by BAR, followed by the bar diameter, a circle with a slash through it, and then the length.
- *Square* – Square bar is specified by BAR, followed by the face width, a square with a slash through it, and then the length.
- *Bar (flat)* – Flat bar is specified by BAR, followed by the wide dimension, the narrow dimension, and then the length.
- *Z-bar* – Z-bar is specified by Z, followed by the flange width, the web depth, the flange and web thickness, and then the length. The flange and web are always the same thickness.

- *Hexagonal* – Hexagonal bar is specified by HEX, followed by the bar thickness (measured across the flats), and then the length.
- *Octagonal* – Octagonal bar is specified by OCT, followed by the bar thickness (measured across the flats), and then the length.

Bar shapes and example specifications for them are shown in *Figure 5*.

### 4.1.4 Angles

An angle is an L-shaped bar. The legs are always at 90 degrees to each other. The legs may be equal-sized or unequal-sized. For equal leg angles, the specification lists the leg width only once followed by its thickness. For unequal leg angles, each leg width is specified followed by its thickness. The last dimension is the length. Angles are shown in *Figure 6*.
Example specifications:

Equal leg angle: L 2" × ¼" × 6'

Unequal leg angle: L 3" × 2" × ¼" × 8'

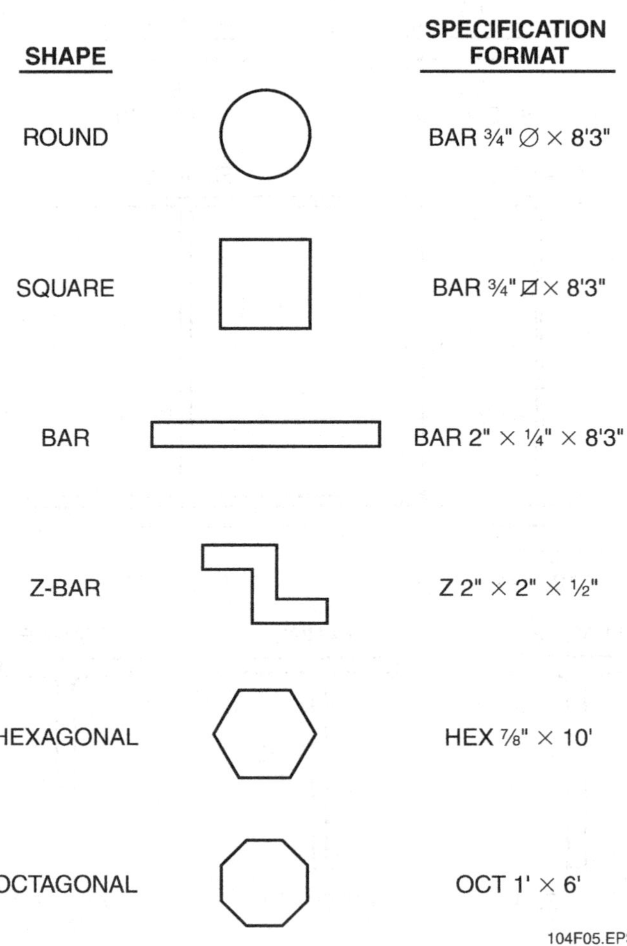


*Figure 5* Bar shapes and specification formats.

### 4.1.5 Channels

Channels are U-shaped forms (*Figure 7*). They are made of two flanges connected by a common web. The flanges extend from the same side of the web. The flanges can be of uniform thickness or tapered toward the outer edges. Channels are classified under two types:

*American Standard channels*
Example specification:

C8" × 11.5 × 20'

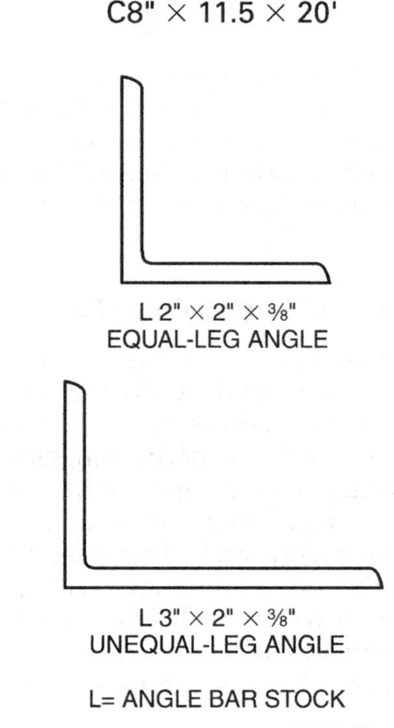


*Figure 6* Angles and specification formats.

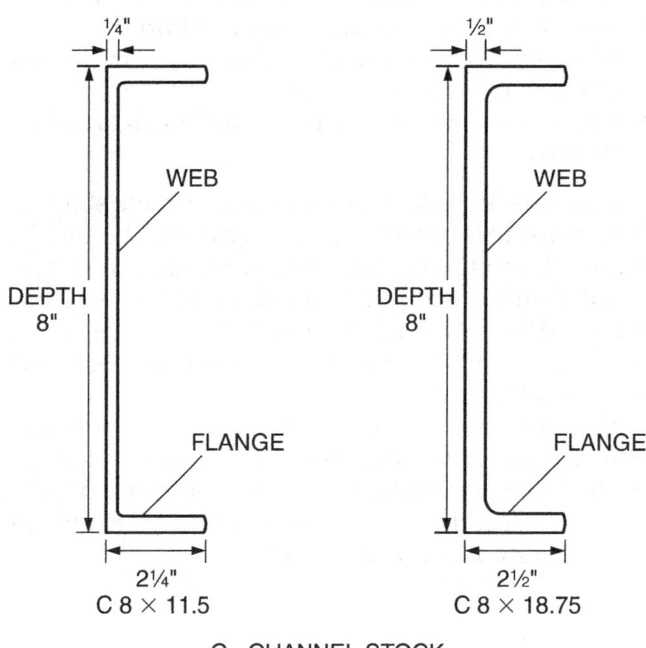


*Figure 7* Channels and specification formats.

The first number is the depth along the web; the second number is the weight in pounds per linear foot; the third number is the length. This method of ordering channel is standard. In some cases, however, the flange width may also be included, so the specification would read C8" × 2¼" × 11.5 × 20', where 2¼" is the flange width.

*Miscellaneous channels*
Example specification:

$$MC6 \times 15 \times 60'$$

The first number is depth in inches along the web. The second number is the weight in pounds per linear foot, and the third number is the length being ordered. Miscellaneous channels are usually lighter than American Standard channels.

### 4.1.6 Beams and Shapes From Beams

Beams are made in I-, H-, T- and Z-shaped cross sections. They are made with flat or tapered flanges. In addition to different flange widths and web depths, the thickness of the flanges and webs varies with beam sizes. Beam weight is measured in pounds per linear foot. Beam weight for a beam of a given dimension can be increased by adding thickness to the web and flanges with very little change in beam width or depth. *Figure 8* shows structural beam and T-beam shapes.

I- and H-beam letter symbols include the following:

* *W* – (Wide flange) wider flanges than S-beams with thinner webs and non-tapered flanges
* *S* – I-shaped beams with tapered flanges
* *M* – An additional classification for beams other than S- or W-beams.
* *HP* – H-shaped beams with non-tapered flanges

T-beams include structural tees and tee shapes. Structural tees are made by cutting S, W, and M beams down the center, usually by shearing. The usual symbol for a structural tee is the original beam letter followed by a T. For example, a tee cut from an S 8 × 18.4 beam would be identified as ST 4 × 9.2.

T-shapes are rolled into their final tee shape. The symbol for tee shapes is a T without any other letter. The specification gives the nominal depth, flange width, thickness, and length. An example specification is T 3 × 3 × 7.8.

### 4.1.7 Steel Pipe

Pipe is listed in inches by its nominal size. For sizes of pipe up to and including 12 inches, nominal size is an approximation of the inside diameter of a Schedule 40 pipe. From 14 inches on, nominal size reflects the outside diameter of the pipe.

The nominal size of a pipe and its actual inside or outside diameter differ greatly, but nominal size is used to describe the pipe. The important thing to remember about size is that the sizes of different types of materials vary. A ¾-inch brass tube fitting will not fit a ¾-inch copper pipe. Brass pipe fittings, however, can be used with threaded steel pipes.

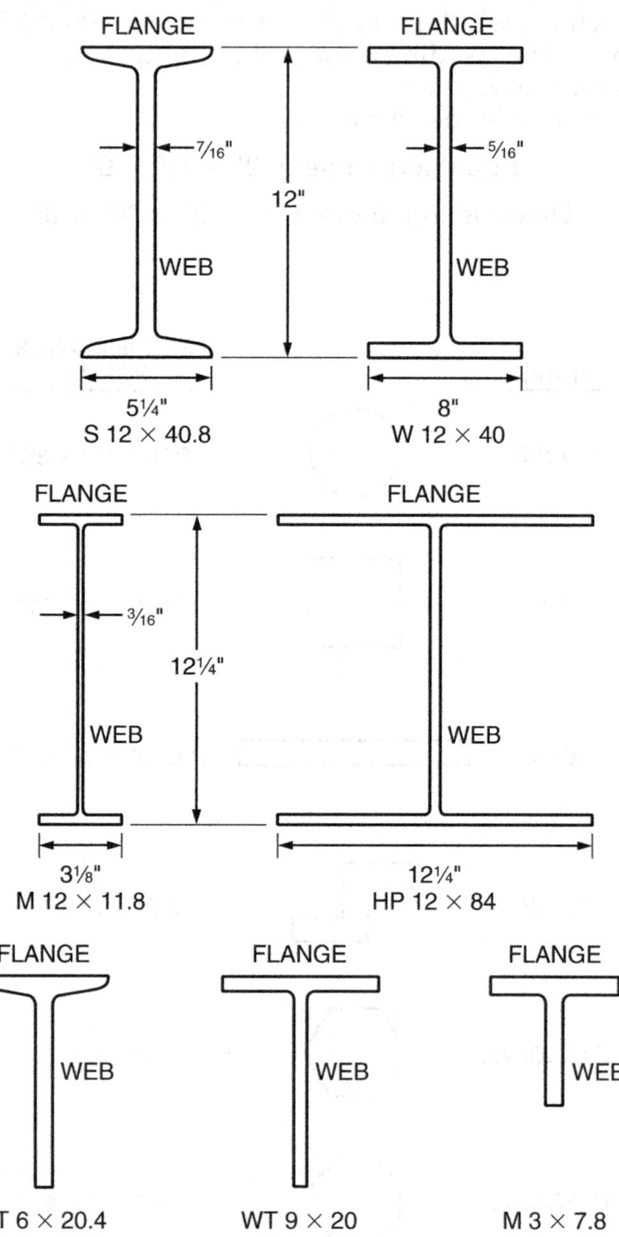


*Figure 8* Structural beam and T-beam shapes.

The wall thickness of pipe can be described in two ways. The first is by schedule. As the schedule numbers increase, the wall thickness gets larger. This means the pipe is stronger and can withstand more pressure. Schedule numbers for pipe range from 5 to 160. No pipe smaller than Schedule 40 should be threaded, although many specifications do not allow threading of pipe smaller than Schedule 80. It is important to remember that the schedule number refers only to the wall thickness of a pipe of a given nominal size. A ¾-inch Schedule 40 pipe will not have the same wall thickness as a 1-inch Schedule 40 pipe. Another way to describe pipe wall thickness is by manufacturer's weight. From smallest to largest, there are three classifications in common use today:

- STD – Standard wall
- XS – Extra-strong wall
- XXS – Double extra-strong wall

Schedule numbers and wall thicknesses are somewhat interchangeable on iron pipe. Schedule 40 galvanized pipe and standard wall galvanized pipe have the same wall thickness for all sizes up to and including 10-inch nominal size. Schedule 80 and extra-strong wall galvanized pipe have the same wall thickness through 8-inch nominal size. Common wall thicknesses of carbon steel pipe are Schedule 40, 80, and 160, standard extra-strong, and double extra-strong. Common wall thicknesses of stainless steel pipe range from Schedule 5 to 160. *Table 7* shows commercial pipe dimensions.

### 4.1.8 Tubing

There are a number of tube materials and many variations of those materials. Tubes can be seamless or welded, cold drawn or heat finished, and annealed or not annealed. There are as many grades as there are alloys of the material they are made of. Tubing manufacturers use standard ASME markings to identify them.

Tubing is manufactured in square, rectangular, and round cross sections. Round tubing can be distinguished from pipe by its dimensions. Standard round tubing has an even-sized OD, ID, and wall thickness.

The cross sections of square and rectangular tubing have even outside dimensions and slightly rounded corners. *Figure 9* shows standard tubing shapes and specification formats.

Pressure tubing used in the construction of boilers is always round. It is identified by its actual outside diameter and wall thickness.

### 4.1.9 Seamed and Seamless Tubing and Pipe

Both standard and non-standard shapes can also be produced by other processes. For example, seamless tubing can be produced by extrusion. Seamed tubing can be produced by rolling and welding. Extrusions can be produced in many shapes.

### 4.1.10 Forged Shapes

Forging is used to produce specialized shapes where high strength is required. An example is forged high-pressure pipe fittings, such as flanges and elbows.

### 4.1.11 Cast Shapes

Cast iron is used in rodding ports, inspection ports, manways, and air ports. It is usually protected by refractory materials. Cast iron can only be welded for emergency repair purposes.

### 4.2.0 Cladding

Cladding (*Figure 10*) is the process of fusing a coating of one type of metal to the surface of another. This allows the use of a base metal with certain desirable features to be combined with the surface feature of another metal. Cladding is usually added to provide corrosion resistance, heat resistance, and wear resistance.

A corrosive environment outside a pipe may indicate the use of stainless steel. Cladding allows the use of low-cost carbon steel. This reduces construction costs. Tube shields are also sometimes used to protect the tubes from wear.

## 5.0.0 INSULATING MATERIALS

Many of a boiler's components normally operate at very high temperatures. These high-temperature items are insulated for three specific reasons:

- To protect personnel working in the area
- To protect equipment operating in the area
- To increase overall efficiency by reducing heat losses to the atmosphere

*Figure 11* shows a typical insulation installation on a boiler wall. Metal lagging is usually installed on exterior boiler walls to prevent damage to insulation.

### 5.1.0 Mineral Wool

Mineral wool is a type of insulation that has been used for many years. Today it may be referred to as rock wool or mineral fiber. Mineral fiber, how-

**Table 7** Standard Pipe Sizes and Wall Thicknesses

| Nominal Pipe Size | Outside Diam. | Nominal Wall Thickness | | | | | | | | | | | | |
|---|---|---|---|---|---|---|---|---|---|---|---|---|---|---|
| | | Sched. 10 | Sched. 20 | Sched. 30 | STD | Sched. 40 | Sched. 60 | XS | Sched. 80 | Sched. 100 | Sched. 120 | Sched. 140 | Sched. 160 | XXS |
| ⅛ | 0.405 | — | — | — | 0.068 | 0.068 | — | 0.095 | 0.095 | — | — | — | — | — |
| ¼ | 0.540 | — | — | — | 0.088 | 0.088 | — | 0.119 | 0.119 | — | — | — | — | — |
| ⅜ | 0.675 | — | — | — | 0.091 | 0.091 | — | 0.126 | 0.126 | — | — | — | — | — |
| ½ | 0.840 | — | — | — | 0.109 | 0.109 | — | 0.147 | 0.147 | — | — | — | 0.188 | 0.294 |
| ¾ | 1.050 | — | — | — | 0.113 | 0.113 | — | 0.154 | 0.154 | — | — | — | 0.219 | 0.308 |
| 1 | 1.315 | — | — | — | 0.133 | 0.133 | — | 0.179 | 0.179 | — | — | — | 0.250 | 0.358 |
| 1¼ | 1.660 | — | — | — | 0.140 | 0.140 | — | 0.191 | 0.191 | — | — | — | 0.250 | 0.382 |
| 1½ | 1.900 | — | — | — | 0.145 | 0.145 | — | 0.200 | 0.200 | — | — | — | 0.281 | 0.400 |
| 2 | 2.375 | — | — | — | 0.154 | 0.154 | — | 0.218 | 0.218 | — | — | — | 0.344 | 0.436 |
| 2½ | 2.875 | — | — | — | 0.203 | 0.203 | — | 0.276 | 0.276 | — | — | — | 0.375 | 0.552 |
| 3 | 3.500 | — | — | — | 0.216 | 0.216 | — | 0.300 | 0.300 | — | — | — | 0.438 | 0.600 |
| 3½ | 4.000 | — | — | — | 0.226 | 0.226 | — | 0.318 | 0.318 | — | — | — | — | — |
| 4 | 4.500 | — | — | — | 0.237 | 0.237 | — | 0.337 | 0.337 | — | 0.438 | — | 0.531 | 0.674 |
| 5 | 5.563 | — | — | — | 0.258 | 0.258 | — | 0.375 | 0.375 | — | 0.500 | — | 0.625 | 0.750 |
| 6 | 6.625 | — | — | — | 0.280 | 0.280 | — | 0.432 | 0.432 | — | 0.562 | — | 0.719 | 0.864 |
| 8 | 8.625 | — | 0.250 | 0.277 | 0.322 | 0.322 | 0.406 | 0.500 | 0.500 | 0.594 | 0.719 | 0.812 | 0.906 | 0.875 |
| 10 | 10.750 | — | 0.250 | 0.307 | 0.365 | 0.365 | 0.500 | 0.500 | 0.594 | 0.719 | 0.844 | 1.000 | 1.125 | 1.000 |
| 12 | 12.750 | — | 0.250 | 0.330 | 0.375 | 0.406 | 0.562 | 0.500 | 0.688 | 0.844 | 1.000 | 1.125 | 1.312 | 1.000 |
| 14 OD | 14.000 | 0.250 | 0.312 | 0.375 | 0.375 | 0.438 | 0.594 | 0.500 | 0.750 | 0.938 | 1.094 | 1.250 | 1.406 | — |
| 16 OD | 16.000 | 0.250 | 0.312 | 0.375 | 0.375 | 0.500 | 0.656 | 0.500 | 0.844 | 1.031 | 1.219 | 1.438 | 1.594 | — |
| 18 OD | 18.000 | 0.250 | 0.312 | 0.438 | 0.375 | 0.562 | 0.750 | 0.500 | 0.938 | 1.156 | 1.375 | 1.562 | 1.781 | — |
| 20 OD | 20.000 | 0.250 | 0.375 | 0.500 | 0.375 | 0.594 | 0.812 | 0.500 | 1.031 | 1.281 | 1.500 | 1.750 | 1.969 | — |
| 22 OD | 22.000 | 0.250 | 0.375 | 0.500 | 0.375 | — | 0.875 | 0.500 | 1.125 | 1.375 | 1.625 | 1.875 | 2.125 | — |
| 24 OD | 24.000 | 0.250 | 0.375 | 0.562 | 0.375 | 0.688 | 0.969 | 0.500 | 1.218 | 1.531 | 1.812 | 2.062 | 2.344 | — |
| 26 OD | 26.000 | 0.312 | 0.500 | — | 0.375 | — | — | 0.500 | — | — | — | — | — | — |
| 28 OD | 28.000 | 0.312 | 0.500 | 0.625 | 0.375 | — | — | 0.500 | — | — | — | — | — | — |
| 30 OD | 30.000 | 0.312 | 0.500 | 0.625 | 0.375 | — | — | 0.500 | — | — | — | — | — | — |
| 32 OD | 32.000 | 0.312 | 0.500 | 0.625 | 0.375 | 0.688 | — | 0.500 | — | — | — | — | — | — |
| 34 OD | 34.000 | 0.312 | 0.500 | 0.625 | 0.375 | 0.688 | — | 0.500 | — | — | — | — | — | — |
| 36 OD | 36.000 | 0.312 | 0.500 | 0.625 | 0.375 | 0.750 | — | 0.500 | — | — | — | — | — | — |
| 42 OD | 42.000 | — | — | — | 0.375 | — | — | 0.500 | — | — | — | — | — | — |

104T07.EPS

ever, is a generic term that includes all common fibrous materials used in insulation. Fiberglass, for example, is mineral fiber.

Mineral wool was one of the first materials used commercially for insulation. It was discovered when escaping steam from a volcano passed through hot lava. This caused mineral fibers to form. Mineral wool was later produced for commercial use by injecting steam into molten slag, the byproduct of steel making.

Modern mineral wool is much like fiberglass, but mineral wool was used years before fiberglass. Generally, mineral wool has a higher temperature limit than fiberglass. It is used for high-temperature applications. Some types used for fireproofing are fire resistant up to 2,000°F. Mineral wool is also used as sound insulation.

Mineral wool is available in many forms. These include preformed pipe insulation and light and heavy density board. Different facings are available. Molded pipe insulation is available with all service jackets (ASJ) and other jackets.


Mineral fiber insulation may be a skin irritant. Long sleeves and loose fitting attire should be worn. If particles or fibers collect on exposed skin areas, do not rub or scratch; instead, wash the area with soap and water.


Refer to the manufacturers' current material safety data sheets for the latest information on precautions to be used when handling their materials.

## 5.2.0 Calcium Silicate

Calcium silicate block is a molded, high-temperature insulation. It is made of calcium silicate and reinforcing fibers. Its upper temperature limit is 1,200°F.

Calcium silicate block is made in one density (approximately 14 pounds per cubic foot). It is available in thicknesses ranging from 1½ to 5 inches, widths from 6 to 18 inches, and a length of 36 inches. It can be stacked to obtain the required thermal protection. It is manufactured in several forms, including flat block, curved radius block, and scored block.


Since calcium silicate insulation is very dusty, appropriate personal protection should be worn when cuts are being made. Calcium silicate is also highly absorbent. It must be kept dry during storage and application.

## 5.3.0 Cellular Glass

Cellular glass (*Figure 12*) is a glass composite foamed under molten conditions. It sets to form a rigid, sealed, cellular insulation. Although its operating temperature ranges from –450°F to 900°F, cellular glass block is used more widely on cold or low-temperature applications.

Cellular glass block weighs about 8.5 pounds per cubic foot. It is fabricated into various lengths, widths, and thicknesses as required, and in flat, curved, or shaped form. Foamed cellular glass is cut with a saw or knife.

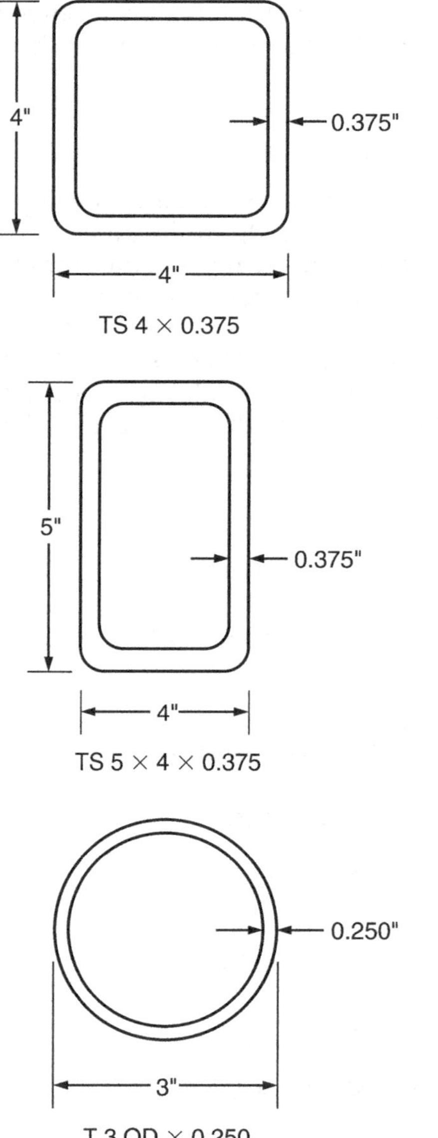


*Figure 9* Standard tubing shapes and specification formats.

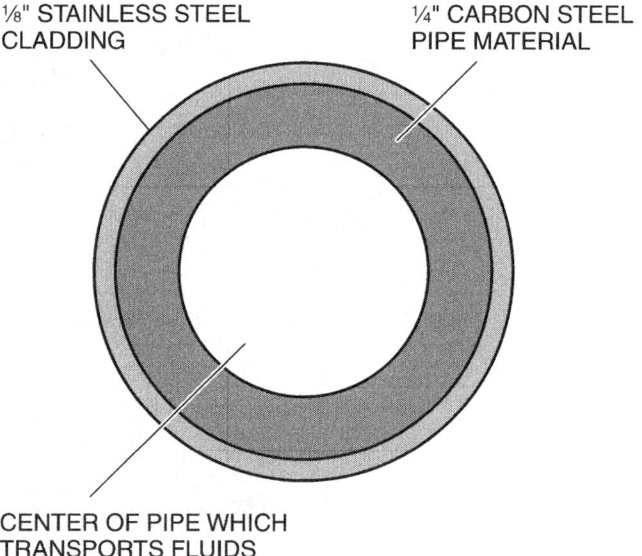


*Figure 10* Cladding example.

## 5.4.0 Fiberglass

Fiberglass is a lower-temperature mineral wool insulation. It is usually used for temperatures up to 900°F and, in some cases, up to 1,200°F. It is available as molded pipe coverings, boards, blankets, and wraps for pipes and vessels. Like mineral wool, it is supplied in various thicknesses. Many products are also available with moisture-proof paperless facings.

## 5.5.0 Perlite

Perlite is a volcanic glass. Perlite insulation is normally supplied as a molded 1,200°F insulation. The insulation is made of an expanded perlite with a sodium silicate binder. The silicate reduces corrosion. Perlite insulation also resists water absorption. It provides excellent thermal insulation. It is available in block or board form and in preformed pipe form along with preformed fitting coverings. Perlite is also used as a material in refractory bricks and mortar.

## 5.6.0 High-Temperature Insulating Cement

High-temperature insulating cement is a mixture of mineral wool or fiber pellets and other inorganic materials. It is used to insulate small-sized pipe fittings and irregular shapes. It is also used to fill voids and cracks in insulation on high-temperature applications. One of the more common trade names for this material is Super #3000®.

High-temperature insulating cement can be used on surfaces that will operate at temperatures up to 1,700°F to 1,900°F. It can be applied to cold or hot surfaces. If high-temperature insulating cement is being applied to a large surface, a layer of poultry mesh is installed for reinforcement. For thicker applications, more than one layer of poultry mesh may be required. It may be used as a one-coat insulation over a large area; however, due to cracking and adherence problems, it is rarely used in this manner.

## 5.7.0 Refractories

Refractory materials, usually simply called refractories (*Figure 13*), are installed inside boilers to withstand the extreme temperatures of combustion. The most common type of refractory is fire brick. Fire bricks are held in place with high-temperature cement. These are not the same bricks and cement used in regular construction. They are specially made to withstand the high temperature

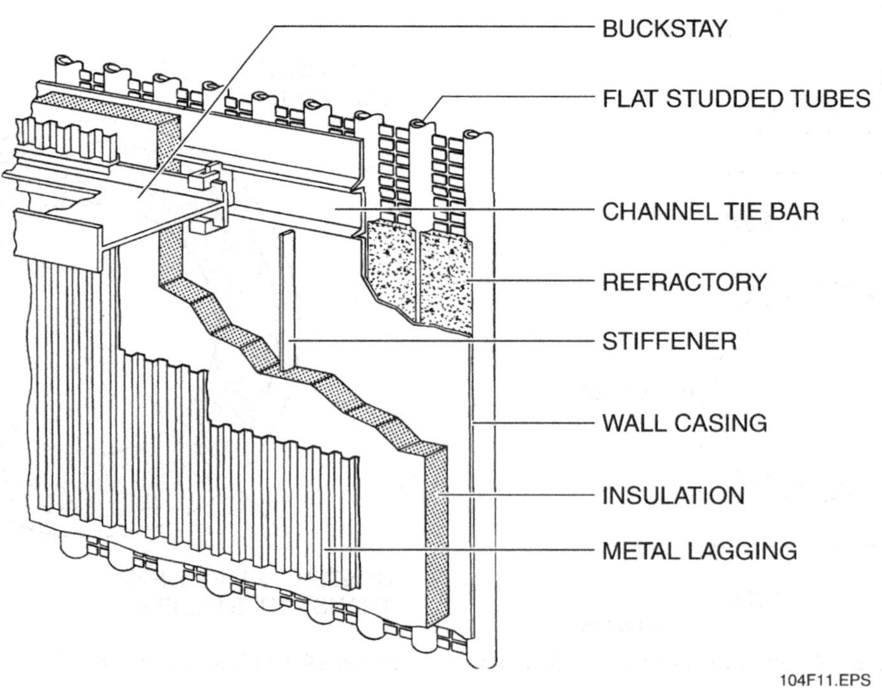


*Figure 11* Boiler wall insulation.

and flame present inside a boiler. Other types of refractories include precast and castable ceramics.

Many areas of a boiler and its associated support equipment must withstand not only extremes of temperature and pressure, but also a large amount of mechanical abrasion. The coal dust used as a fuel in a coal-fired boiler, for example, acts much like the abrasive compound used during sandblasting when fed to the boiler components during operation.

In order to withstand these conditions, critical areas of a boiler are often covered with ceramic materials that withstand high heat and mechanical wear. These materials must be installed by specially trained personnel. Some ceramics are custom-formed shapes or tiles. Others are site-applied monolithic castables. Of the castables, some are dense and noninsulating. Others are insulat-

ing, and still others are abrasion-resistant. These ceramics may be applied to forms. They may be pumpable or gunning mixes applied to surfaces.

Ceramics are common in boilers that burn trash. The fuels used in these boilers contain common household and industrial trash. Such fuels may include abrasive material. Ceramics are used heavily in the furnace area of this type of boiler to withstand the abrasion. They are also used where the velocity of the exhaust gases is increased due to decreased size. They are used in ash hoppers as well.

104F12.EPS

*Figure 12* Cellular glass products.

104F13.EPS

*Figure 13* Refractories.

## Summary

Boilers are complex devices, made of many different materials. Boilermakers must understand the many properties of the materials used in boiler work, so that they can select the appropriate material to use for any tasks that they may be called on to perform. Improper materials can cause unnecessary expense, loss of efficiency, premature equipment failure, and even dangerous ruptures or explosions. For this reason, a good understanding of materials is essential for the boilermaker.

1. A substance composed of two or more elements, chemically joined, is known as a(n) _____.

   a. compound
   b. ion
   c. mixture
   d. alloy

2. The heat transfer ability of a substance is also known as its _____.

   a. electrical conductivity
   b. thermal conductivity
   c. density
   d. thermal expansion rate

3. The ability of a material to undergo repeating stress or load is known as its _____.

   a. thermal expansion rate
   b. tensile strength
   c. fatigue strength
   d. melting point

4. In the SAE International system of steel designation, 4040 designates a steel that contains _____.

   a. 0.4 percent carbon
   b. 0.8 percent carbon
   c. 40 percent carbon
   d. 80 percent carbon

5. In the SAE International system of steel designation, the first two numbers indicate the metal's _____.

   a. alloy content
   b. carbon content
   c. tensile strength
   d. ductility

6. Adding nickel to steel increases its _____.

   a. ductility
   b. corrosion resistance
   c. malleability
   d. strength and toughness

7. When specifying the dimensions of angle iron, the last digit indicates the material's _____.

   a. leg width
   b. length
   c. angle
   d. thickness

8. Schedule numbers for pipe products are a measure of their _____.

   a. length
   b. outside diameter
   c. inside diameter
   d. wall thickness

9. Cladding is placed around a metallic product in order to increase its _____.

   a. corrosion resistance
   b. ability to be welded
   c. tensile strength
   d. ductility

10. Ceramics are used in the construction of trash-burning boilers primarily to _____.

    a. reduce the cool-down time
    b. decrease the overall weight of the boiler
    c. increase wear resistance
    d. increase the ability to withstand high pressures

## Roger Ivey
KBR Construction Co., Inc.
Regional Manager

*How did you choose a career in the boilermaking field?*
I originally started my career in the layout and fabrication industry. It consisted of plate, structural, and pipe fabrication. To advance my resume, I became involved in the erection process.

*Who inspired you to enter the industry?*
No one really inspired me; I was just lucky to become employed into the trade. In the early 60s, you were very lucky to find employment in this field (normally if someone retired).

*How important is education and training in construction?*
In the fast-paced world of boilermaking today, it is difficult for one-on-one training, so any training you can obtain in your trade is very important.

*How important are NCCER credentials to your career?*
Throughout my career, I was able to obtain direct on-the-job training by very competent professionals. However, I recognize the importance of NCCER credentials, and I strive to emphasize this to workers that are progressing in the boilermaking trade.

*How has training/construction impacted your life?*
The training that has had the greatest impact on me personally has been safety training, as I started when there was not much safety training for the industry.

*What kinds of work have you done in your career?*
Over a span of 50 years, I have worked in all aspects of the fabrication field, including boiler erection, piping fabrication and erection, steel erection, rigging, equipment operation, equipment setting, and civil. I have been in supervision, planning, and the scheduling of listed trades.

*Tell us about your present job.*
I am now a district manager for BE&K Boiler group, the southwest region.

*What do you enjoy most about your job?*
Working with people and being able to look at a finished project and knowing you were involved with accomplishing the end results.

*What factors have contributed most to your success?*
As a young man I was lucky that some highly experienced tradesmen recognized my work ethics and the desire to learn. These people took me under their wings and shared all their knowledge to help me advance. This was back when you had to learn it to earn it. Raises came a nickel and dime at a time and all advancements were demonstrated hands on.

*Would you suggest construction as a career to others? Why?*
Yes. It is a rewarding trade and one with a lot of challenges. It is also a very profitable trade, with rewards of a great financial future.

*What advice would you give to those new to boilermaking field?*
Align yourself with the most experienced people and do not be afraid to ask questions. Show your willingness to be a good employee for the future. Become involved in as many training programs as possible, especially NCCER.

*Interesting career-related fact or accomplishment:*
Going to China as an erection manager of two boilers, two precipitators, and turbines. Being able to see just how far we have come as a nation compared to others in the construction industry.

*How do you define craftsmanship?*
Craftsmanship is the individual recognizing that if it is worth doing it is worth doing right the first time. It is also having pride in knowing that you will be able to deliver a finished product others will recognize as craftsmanship.

# Trade Terms Introduced in This Module

**American Iron and Steel Institute (AISI):** An industry organization responsible for preparing standards for steels and steel alloys based upon the SAE code system.

**Alloy:** A fused mixture of two or more metals, or a metal and one or more other metals and/or non-metals.

**ASTM International:** An organization that developed a code system for identifying and labeling steels.

**American Society of Mechanical Engineers (ASME):** An organization that has developed and maintains many of the codes used in the construction of boilers and pressure vessels.

**Coefficient of thermal expansion:** The amount a unit length of material expands for each degree of temperature increase.

**Ductility:** A metal's ability to be drawn, stretched, or hammered thin without breaking.

**Ferrous:** Relating to iron or an alloy that contains mostly iron.

**Malleability:** A metal's ability to be hammered or pressed into another shape without breaking.

**SAE International:** An automotive society responsible for an early system of classifying carbon steels.

**Steel:** Iron alloyed with carbon and/or other metals and elements to produce specific characteristics of toughness, hardness, and corrosion resistance.

**Unified Numbering System (UNS):** A designation system used in North America only for commercial metals and alloys that are in active use.

**Wrought:** Formed or shaped by hammering or rolling.

**Yield stress:** The amount of pulling stress which can be placed on a material before permanent deformation occurs. It is one measure of a material's strength.

## Additional Resources

This module is intended to present thorough resources for task training. The following references are suggested for further study. These are optional materials for continued education rather than for task training.

*Metallurgy for the Non-Metallurgist*, Latest Edition. Materials Park, OH: ASM International.

*Welding Level One Trainee Guide*, Latest Edition. Upper Saddle River, NJ: Prentice Hall.

*Worldwide Guide to Equivalent Irons and Steels*, Latest Edition. Materials Park, OH: ASM International.


## Figure Credits

© iStockphoto.com/Michael Westhoff, Module opener

Pannier Corporation, 104F03, 104F04

The Babcock & Wilcox Company, 104F11

Pittsburgh Corning Corporation, 104F12

BNZ Materials, Inc., 104F13